Shoals and Schools of Fish

Rachel Stuckey

A Crabtree Crown Book

Crabtree Publishing
crabtreebooks.com

Author: Rachel Stuckey

Series research and development:
Reagan Miller
Janine Deschenes

Editor: Janine Deschenes

Proofreader: Kathy Middleton

Design and photo research:
Katherine Berti

Print and production coordinator:
Katherine Berti

Image credits:
Alamy Stock Photo
Hudson Fleece: p. 20–21
iStockphoto
AG: p. 14 (bottom)
Wikimedia Commons
USFWS: p. 13 (bottom)
All other images by Shutterstock

Crabtree Publishing

crabtreebooks.com 800-387-7650

 In Canada: We acknowledge the financial support of the Government of Canada through the Canada Book Fund for our publishing activities.

Hardcover	978-1-0398-0655-9
Paperback	978-1-0398-0681-8
Ebook (pdf)	978-1-0398-0706-8
Epub	978-1-0398-0733-4

Published in Canada
Crabtree Publishing
616 Welland Avenue
St. Catharines, Ontario
L2M 5V6

Published in the United States
Crabtree Publishing
347 Fifth Avenue
Suite 1402-145
New York, New York, 10016

Library and Archives Canada Cataloguing in Publication
Available at Library and Archives Canada

Library of Congress Cataloging-in-Publication Data
Available at the Library of Congress

Printed in the U.S.A./012023/CG20220815

Contents

What Are Shoals and Schools?

Think of a time that you saw a fish in the water. Was it alone, or did you see many other fish, too? Fish group together for different reasons. They might group together naturally, to feed on the same thing or hide in the same place. But when fish choose to group together, it is called a shoal or a school.

There are thousands of different types of fish. Some live in lakes and rivers. Others live in oceans.

When fish gather in a loose cluster, it is called shoaling. In a shoal, the fish do different things and face different directions. When fish swim together in a **coordinated** way, it is called schooling. Schools of fish swim in a form or shape. They also move together at the same time.

Fish can switch between shoaling and schooling. When fish stop swimming in a school, they become a shoal.

Most rainbowfish species are schooling fish.

Did You Know?

About one quarter of the world's fish **species** are schooling fish. These are fish that spend most of their lives in shoals and schools. Other species of fish may form shoals or schools only at certain times or for a certain reason.

How Do Fish Swim in Schools?

Scientists think that fish might swim better in a school.

Schools can be different sizes and shapes. In the ocean, thousands of fish could form a school. In an aquarium, the smallest schools are made up of four to six fish. Schools may take the shape of thin lines, squares, or ovals. When schools swim very fast, they often form a wedge shape.

When schooling, fish use their eyes to watch each other. The fish can also feel **vibrations** in the water. The vibrations tell them the locations of the other fish in the school. If one fish moves closer, the other fish moves away. This is how the fish avoid running into each other.

Sticking Together

There is no leader in a school of fish. But some fish will tend to swim at the edges of the school and other fish will tend to swim in the middle. Some studies show that fish rotate positions in the school.

Why Do Fish Shoal and School?

There are three main reasons that fish shoal and school together. First, shoals and schools give fish safety. In a group, there are more fish to watch for predators. Schools of fish can confuse predators, too. As they swim together, they form balls that shimmer, swirl, and change direction quickly.

When groups of fish swim close together, the predator cannot see a single fish to chase.

Second, there are more eyes to look for food in shoals and schools. When a school of fish stops to feed, it becomes a shoal. Third, fish in shoals and schools have an easier time finding others to **mate** with. This is because most groups already include both male and female fish.

In habitats with less food, fish are more likely to look for food on their own. That way, they don't have to share the food they find with other fish.

Did You Know?

Young fish are called juveniles. They grow in stages. In the first stage, they are known as fry. As they grow bigger, they are called fingerlings. Teenage fish are called smolts.

Freshwater Fish

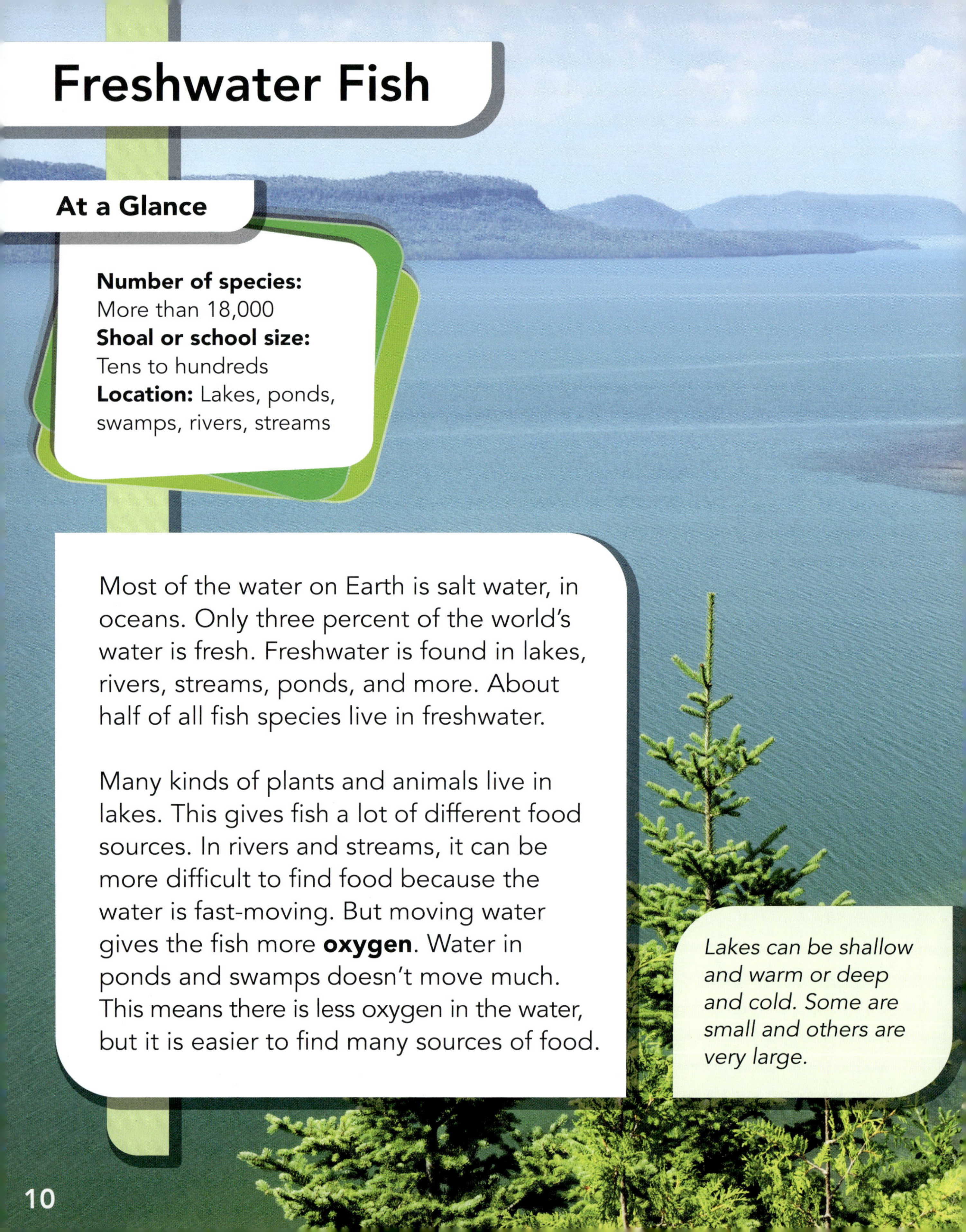

At a Glance

Number of species: More than 18,000
Shoal or school size: Tens to hundreds
Location: Lakes, ponds, swamps, rivers, streams

Most of the water on Earth is salt water, in oceans. Only three percent of the world's water is fresh. Freshwater is found in lakes, rivers, streams, ponds, and more. About half of all fish species live in freshwater.

Many kinds of plants and animals live in lakes. This gives fish a lot of different food sources. In rivers and streams, it can be more difficult to find food because the water is fast-moving. But moving water gives the fish more **oxygen**. Water in ponds and swamps doesn't move much. This means there is less oxygen in the water, but it is easier to find many sources of food.

Lakes can be shallow and warm or deep and cold. Some are small and others are very large.

Freshwater fish can also live in lakes and ponds built by people. Fish that live in human-made ponds are safer from predators. They also have plenty of food.

trout

Many fish species, such as trout and perch, thrive in lakes.

perch

Yellow Perch

At a Glance

Shoal or school size: 50 to hundreds
Location: Lakes in North America

Yellow perch school in deep, cool water during the day when it is warm. They move to shallow waters to feed at dusk and dawn when the temperature goes down. Smaller perch swim in the middle of the school and larger perch are on the edge. This gives the smaller fish extra protection and the bigger fish better access to food.

Yellow perch are yellow gold in color. They have thick, vertical stripes and a humpback.

White Bass

At a Glance

Shoal or school size: Hundreds
Location: Rivers, creeks, streams, and lakes in North America

The white bass is about 12 inches (30 cm) long. It has thin stripes along its body.

White bass are schooling fish. They feed in large groups in early morning and late evening. Large schools of white bass work together to catch **prey**. They drive or chase smaller fish to the surface of the lake. The prey fish often leap from the water as they try to escape.

Sticking Together

White bass often school with yellow bass. These fish are a similar size. Usually, schools and shoals include fish that are similar or the same in size, color, and species. This helps protect the group from predators. If one fish sticks out from thc crowd, they are easier for a predator to see. That can put an entire shoal in danger.

Bluegills

At a Glance

Shoal or school size: 10 to 20 fish, schools with other sunfish
Location: Freshwater lakes, rivers, streams east of Rocky Mountains

Bluegills are a type of sunfish. They usually school with just 10 to 20 fish. Shoals of bluegills lay their eggs together in **nesting colonies**. Bluegills also work together to chase predators, such as snapping turtles, away from their eggs. This is called mobbing.

Sometimes, bluegills school with other types of sunfish of the same size. One example is the smallmouth bass.

Piranhas mostly eat smaller fish, insects, worms, and plants. Sometimes, schools of piranha will eat animals that fall into the river.

Piranhas

At a Glance

Number of species: More than 60
Shoal or school size: Usually around 20, but up to 100
Location: Rivers and lakes in South America

Piranhas are a freshwater fish. They are famous as predator fish because of their large teeth. Scientists used to think piranhas swam in schools to hunt prey. But, like most fish, piranhas shoal and school for protection against predators, such as cormorants and caimans.

caiman

cormorant

Marine Fish

At a Glance

Number of species: More than 20,000
Shoal or school size: Tens to hundreds of millions
Location: All oceans

Oceans and seas have salt water. Fish that live in salt water are called marine fish. Even though 97% of the water on Earth is salt water, only half of fish species are marine fish. The ocean is also home to other types of animals, such as dolphins and octopuses.

Oceans have many different types of habitats. For example, there is deep water, surface water, shallow water, and coral reefs. The temperature changes in different parts of the ocean. Some places in the ocean have more sources of food. Different kinds of fish live in different ocean habitats.

Brightly colored blue tangs are one type of marine fish that form schools.

Clownfish live in shallow coral reef habitats. Though they are not schooling fish, they do live in groups.

Bluefin tuna live in parts of the ocean deeper than coral reefs. They travel in large schools.

Did You Know?

Dolphins and whales swim together in groups in the ocean. Sometimes, people call these groups schools. But dolphins and whales are **mammals**, not fish. They do not swim in a coordinated way. Instead, groups of dolphins and whales are called pods.

Groups in the Ocean

Protection, finding food, and **spawning** are some of the reasons why marine fish shoal and school. Schools of ocean fish range in size from just a few to millions of fish.

Small fish, such as anchovies, sardines, and herring, search for food such as **plankton**. These small fish spend most of their lives in schools in the open ocean.

Larger fish, such as cod, hake, and pollack, hunt the smaller fish. They are also prey for predators such as tuna and sharks. Some of the ocean's biggest predators, such as tuna, swim in schools to **migrate** and hunt.

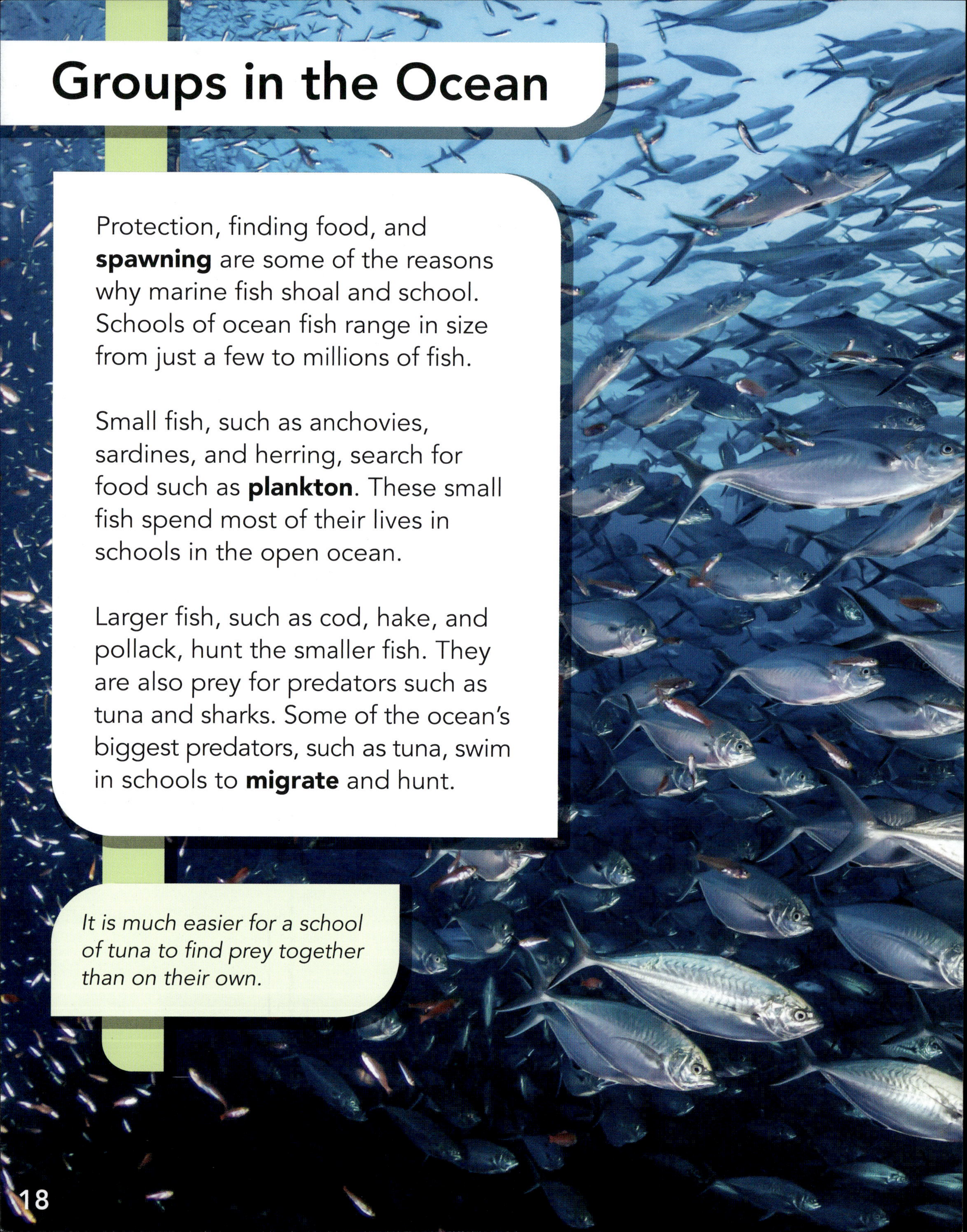

It is much easier for a school of tuna to find prey together than on their own.

Sticking Together

The Humboldt squid is a large **mollusk** that lives in the eastern Pacific Ocean. Because they are not fish, Humboldts cannot swim in a school. However, they do hunt in groups of over 1,000 squid. Groups of Humboldt squid are often called shoals.

Humboldt squid can grow to be 5 feet (1.5 m).

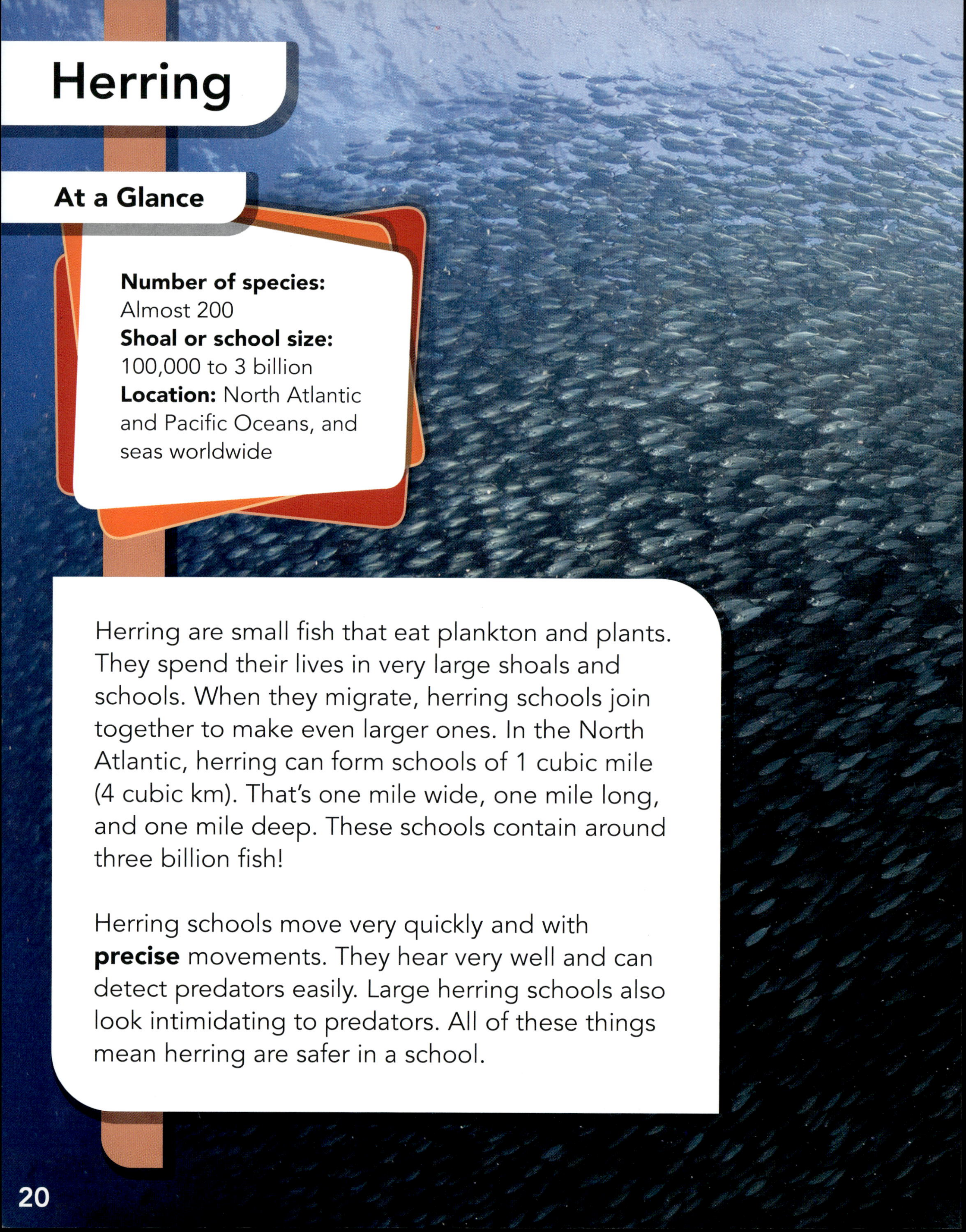

Herring

At a Glance

Number of species: Almost 200
Shoal or school size: 100,000 to 3 billion
Location: North Atlantic and Pacific Oceans, and seas worldwide

Herring are small fish that eat plankton and plants. They spend their lives in very large shoals and schools. When they migrate, herring schools join together to make even larger ones. In the North Atlantic, herring can form schools of 1 cubic mile (4 cubic km). That's one mile wide, one mile long, and one mile deep. These schools contain around three billion fish!

Herring schools move very quickly and with **precise** movements. They hear very well and can detect predators easily. Large herring schools also look intimidating to predators. All of these things mean herring are safer in a school.

Sticking Together

Herring thrive in large groups. When herring are put in aquariums on their own, they become very agitated. Even small groups in aquariums become weak and slower.

Most adult herring are around one foot (30 cm) long.

Ocean Predators

Many species of predator fish also shoal and school. Large predators, such as marlin and tuna, can spot prey faster in a school than on their own. Predator schools will surround a group of prey fish, such as mackerel. Each predator will take turns swimming into the school or shoal to break it up. This makes it easier for the school of predators to catch individual prey fish.

Many species of shark also stick together in shoals. These shoals are often called schools, but they do not swim in a coordinated way.

Schools of predators will swim thousands of miles across the ocean looking for prey.

marlin

Tuna

At a Glance

Number of species: 15
Shoal or school size: 5 to 1200
Location: All oceans

Tuna, a large predator species, have a strong schooling instinct. Different species of tuna will school together if they are of similar size. Schools travel the ocean and hunt prey together. In winter, schools of tuna spawn in tropical oceans. In summer, they swim north to find food.

Tuna are very large. The largest species, Atlantic bluefin tuna, weighs 550 pounds (250 kg) and grows 6.5 feet (2 m), on average.

Shoals and Schools on Tropical Coral Reefs

Coral reefs are home to thousands of marine plant and animal species. Small fish living on coral reefs sometimes form shoals and schools to protect themselves from predators. The yellow goatfish feeds at night by digging through the sand. During the day, the goatfish will school with the blue-striped snapper for protection.

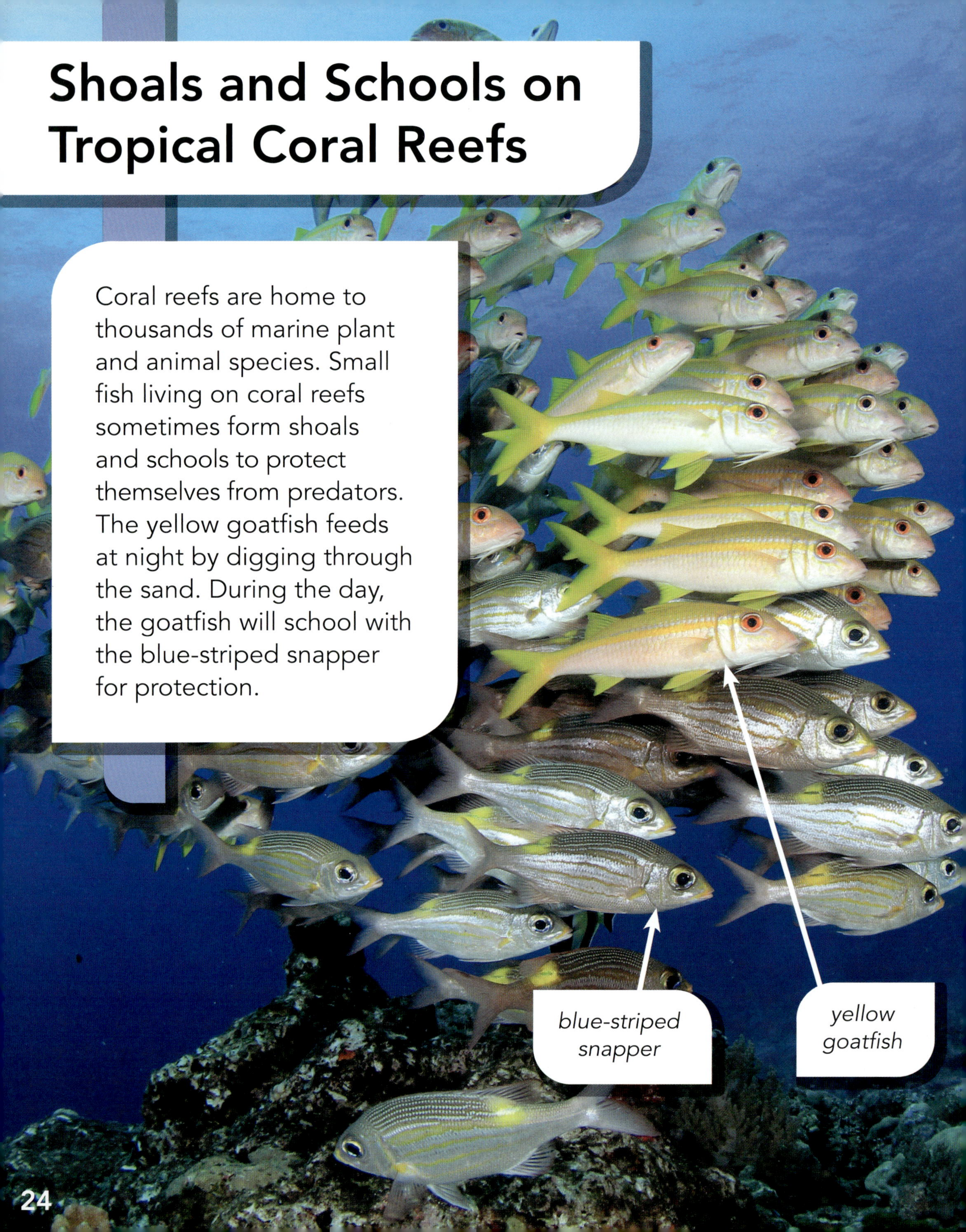

Predators also hunt in groups near coral reefs. Schools of bigeye trevally surround schools of smaller fish. This makes the fish panic and breaks up the group. Then, it is easier for the trevally fish to attack. Some species of barracuda hunt in small schools around coral reefs. The barracuda school swims slowly as it looks for prey. Then it speeds up to chase and catch the prey.

Did You Know?

Coral reefs are created by animals called coral. One type of coral has a hard shell. These coral live in colonies. New colonies grow on top of old ones, forming a coral reef. Over time, it grows larger and larger.

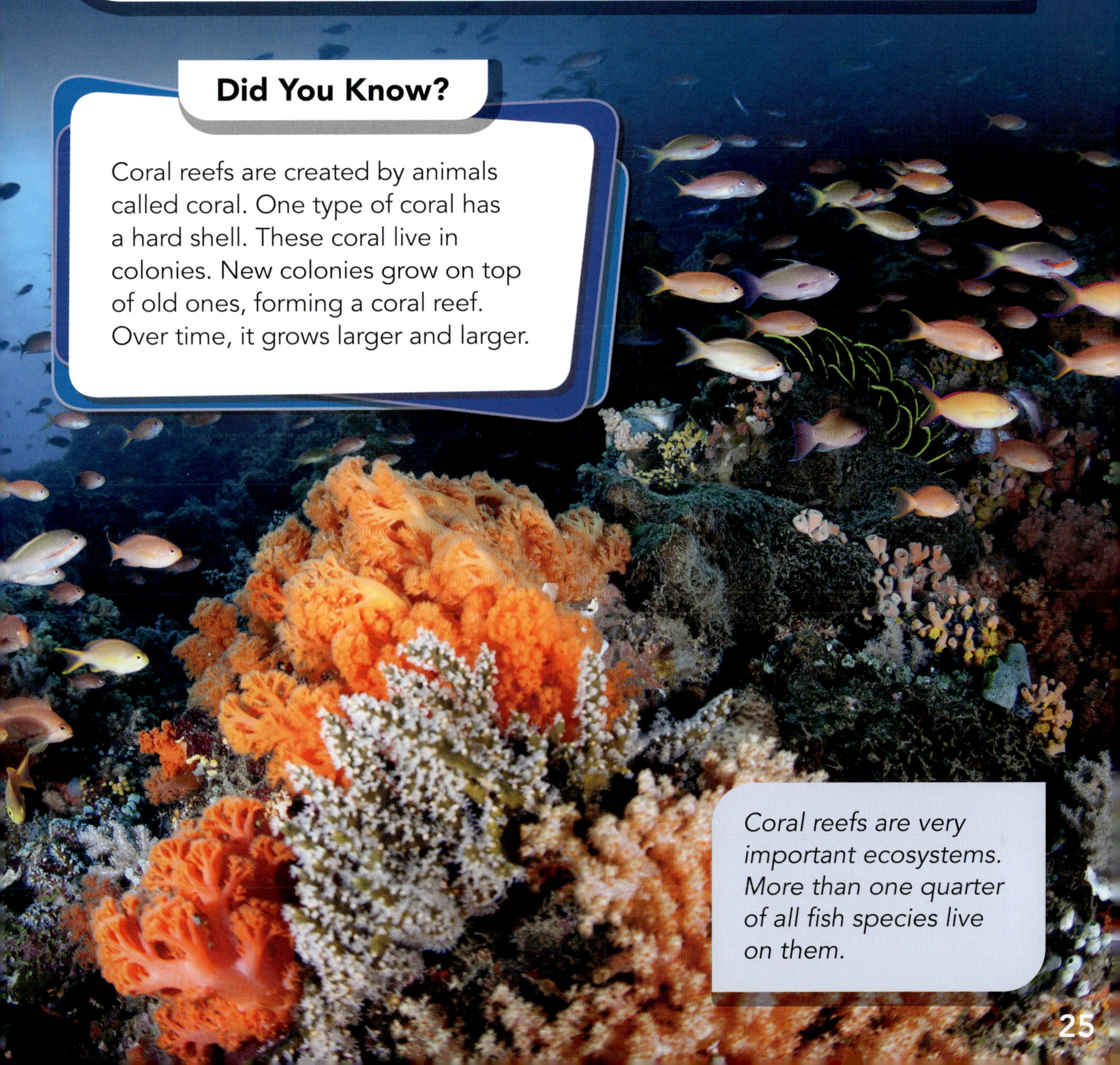

Coral reefs are very important ecosystems. More than one quarter of all fish species live on them.

Rivers to Oceans and Oceans to Rivers

Some fish can live in both freshwater and salt water. These fish migrate from one habitat to the other. Some of these fish will shoal and school, especially when they are in the ocean. But often, they only group together because they migrate at the same time.

Salmon are born in rivers, then swim out to the sea. When they are in the ocean, juvenile salmon school for protection. But adult salmon are **solitary** fish. At the end of its life, a salmon will swim back upriver to spawn. Hundreds of salmon swim upriver at the same time. But this is not a shoal or school. The salmon group together because of their migration.

A school of salmon forms a shiny, shimmery ball that can look like a large predator.

In its **life cycle**, the European eel migrates from the Atlantic Ocean, up rivers, and back to the ocean. It migrates at the same time as other eels. This looks like shoaling or schooling, but eels are solitary fish.

Salmon return to the place where they were born to spawn. All species of Pacific salmon die after spawning. There are some species of Atlantic salmon, however, that can spawn for several years.

Through the Life Cycle

Some fish shoal and school for their entire lives. Other fish do so only at certain times or in certain conditions. Many fish shoal and school when they are young. Juvenile striped eel catfish form ball-shaped schools of about a hundred fish. This helps protect them from predators. As adults, striped eel catfish become solitary. They may form smaller shoals or schools of about 20 fish.

Striped eel catfish search for food in the sand on the floor of the Indian Ocean. This is much easier to do alone.

Largemouth bass are found in lakes in North America.

Largemouth bass spend most of the year in shoals and schools. When it's time to spawn, they become **territorial** and stay away from other fish. This is how they protect their eggs. Rainbow trout are the opposite. They spend their lives as solitary fish until it's time to spawn. Then they swim together in schools. This makes it easier to find a mate.

Rainbow trout live in the North Pacific Ocean and the rivers and lakes that drain into it.

Glossary

coordinated Organized, so many parts work together

life cycle The series of changes that a living thing goes through during its life

mammals Warm-blooded animals with backbones. Mammal babies drink milk from their mothers' bodies.

mate Join together to breed, or have babies

migrate Of animals, move from one place to another, often according to the season

mollusk Animals without backbones, such as octopus and snails

nesting colonies Groups of animals that come together to build nests and lay eggs or have babies

oxygen A gas in the environment that living things need to breathe

plankton Small living things that float or drift in the ocean

precise Very accurate or exact

predators Animals that hunt other animals for food

prey Animals that are hunted by other animals for food

solitary To be alone

spawning Of fish, laying many small eggs

species A group of living things with similar characteristics, in which two animals can have babies

territorial Defending a territory

vibrations Rapid back-and-forth motions

vulnerable Able to be easily hurt